Common Weeds of Canada

Published by
NC Press Limited
in co-operation with
Agriculture Canada
and the
Canadian Government Publishing Centre
Supply and Services Canada

Gerald A. Mulligan

COMMON WEEDS of CANADA

FIELD GUIDE

Second printing, 1989

Catalogue number: A22-83/1986E

Canadian Cataloguing in Publication Data

Mulligan, Gerald A., 1928-
Common weeds of Canada

Co-published by Agriculture Canada.
Includes index.
ISBN 0-920053-59-9

1. Weeds – Canada. 2. Weeds – Canada - Identification. I. Canada. Agriculture Canada. II. Title.

SB613.C3M84 1987 581.6'52'0971 C87-094307-3

We would like to thank the Ontario Arts Council, the Canada Council, and the Department of Communications, Canada for their assistance in the production of this book.

New Canada Publications, a division of NC Press Limited, Box 4010, Station A, Toronto, Ontario M5H 1H8

Printed and bound in Canada

To Marg
thanks for your
support

Contents

Introduction

Weeds are the most prominent plants of settled areas of Canada. We see them every day during the growing season. In cities, weeds inhabit lawns, gardens, and waste places. In the country, they line every roadside and grow vigorously in all fields that have recently been disturbed. We need to know the identity of these plants not only as an aid in weed control but also to satisfy our increasing curiosity about the natural things around us.

By definition, a weed is a plant that grows where man does not want it to grow: in grainfields, row crops, pastures, hayfields, lawns, and other man-made habitats. However, it is doubtful if many of the plants we call weeds could survive or even grow in their present abundance if these artificial habitats did not exist. We are, in fact, largely responsible for creating a suitable environment for the growth of those plants that we are most anxious to eliminate. It is probably impossible to eliminate weeds permanently from any habitat without making that habitat unsuitable for plant growth. Weed control is, therefore, a never-ending battle.

This publication contains 125 coloured illustrations of the most common weeds of Canada. One hundred and nineteen of these illustrations are reproduced from coloured photographs taken by the author. The remaining 8 illustrations are the late Norman Criddle's drawings and are taken from the out-of-print publication *Farm Weeds of Canada.*

Use this publication to identify weeds by visually comparing wild plants with colored plates. Detailed descriptions of weeds are given in the revised edition of *Weeds of Canada* which can be bought from many bookstores.

Field Horsetail *Equisetum arvense* L.

Perennial, spreading by spores and by creeping rootstocks; not a flowering plant; stems 10 to 25 cm high; native to Canada; a common weed in all provinces; occurs in a wide range of habitats; a poisonous plant particularly affecting young horses; unbranched stems ending in a cone appear by the middle of April and soon wither; the green branched stems shown in the photograph appear early in May and last until frost.

Quack Grass *Agropyron repens* (L.) Beauv.

Perennial, spreading by seeds and underground rootstocks; stems 0.35 to 1.35 m high; introduced from Europe; common in settled areas of all provinces; in grasslands, cultivated fields, waste places, and along roadsides; one of the most difficult weeds to control; quack grass is considered one of the world's 10 most serious weeds; flowers appear toward the end of June.

Wild Oats *Avena fatua* L.

Annual, spreading by seeds; stems 0.70 to 1.35 m high; introduced from Europe and Asia; in all provinces; probably the most troublesome weed growing in grainfields of the Prairie Provinces; delayed germination permits the plant to escape fall cultivation and survive the winter as a seed; flowers appear at the beginning of July.

Downy Brome *Bromus tectorum* L.

Annual or winter annual, spreading by seeds; often in large tufts; stems up to 0.7 m high; introduced from Europe; occurs from Nova Scotia to British Columbia; most abundant under dry conditions, particularly in southwestern Alberta and the interior of British Columbia; inconspicuous flowers appear early and seed ripens from June to August.

Smooth Crab Grass *Digitaria ischaemum* (Schreb.) Muhl.

Annual, spreading by seeds; native to Europe and Asia; in all provinces except Newfoundland; aggressive weed of lawns, gardens, waste places, and along roadsides in southern Ontario and Quebec; particularly troublesome in lawns, where its flowering stems, tendency to sprawl, purple color, and rapid growth late in the summer are obnoxious; flowers appear from mid-July to late September.

Barnyard Grass *Echinochloa crusgalli* (L.) Beauv.

Annual, spreading by seeds; stems 0.35 to 1.35 m high; introduced from Europe; common in all eastern provinces, but rare west of Manitoba; in cultivated fields, gardens, barnyards, waste places, ditches, and along roadsides; flowers from July to September.

Foxtail Barley *Hordeum jubatum* L.

Perennial, spreading by seeds; stems 35 to 70 cm high; native to western North America; in every province; in native habitats and in meadows, lawns, waste places, and along roadsides; structures enclosing seed have sharp bristles that can cause serious injury to mouths or skin of livestock; flowers appear in July.

Witch Grass *Panicum capillare* L.

Annual, spreading by seeds; stems a few centimetres to 1 m high; native to North America; in all provinces except Newfoundland; particularly troublesome in gardens and cultivated fields of southern Ontario and southern Quebec; witch grass germinates late, but grows vigorously and is well developed by July; flowers from July to September.

Green Foxtail *Setaria viridis* (L.) Beauv.

Annual, spreading by seeds; stems 8 to 100 cm high; introduced from Europe; in all provinces; particularly troublesome in the Prairie Provinces; in grainfields, gardens, waste places, and along roadsides; seeds of green foxtail germinate mainly between 15 May and 15 June, so that early spring and late summer cultivation have little effect on control; flowers appear from July to September.

Yellow Nut Sedge *Cyperus esculentus* L.

Perennial with rhizomes terminating in tubers or leafy plants; spreading mostly by overwintering tubers; stems triangular, 35 to 75 cm high; native to North America; in Nova Scotia, New Brunswick, southern Quebec, and southern Ontario; in spring-flooded native habitats and in cultivated fields; flowering early July to September.

Stinging Nettle *Urtica dioica* L.

Perennial; stems 0.35 to 2.7 m high; native to North America; in every province; along roadsides, rivers, and streams, and in waste places; contact with stinging hairs can cause great discomfort, an intense itching of short duration; flowers from June to September.

Striate Knotweed *Polygonum achoreum* Blake

Annual, spreading by seeds; prostrate to semi-erect; native to North America; in all provinces except Newfoundland and Prince Edward Island; common along roadsides and in waste places; inconspicuous flowers present from July to September.

Prostrate Knotweed *Polygonum aviculare* L.

Annual, seeds germinate very early in the spring; prostrate to semi-erect; introduced from Europe and Asia; in all provinces; most common on trampled land around habitations, also in waste places, cultivated fields, and along roadsides; flowers from June to September.

Wild Buckwheat *Polygonum convolvulus* L.

Annual, spreading by seeds; introduced from Europe; in settled areas of all provinces; most abundant in the Prairie Provinces; in grainfields, row crops, gardens, waste places, and along roadsides; stems of this weed frequently twine around other plants; flowers in late June and July.

Pale Smartweed *Polygonum lapathifolium* L.

Annual, spreading by seeds; stems 0.35 to 2 m high; native to North America and Europe; in all provinces, but most abundant in the Prairie Provinces; in grainfields, waste places, and along roadsides; its elongate and nodding spikes distinguish it from other smartweeds; flowers first appear in July.

Lady's-Thumb *Polygonum persicaria* L.

Annual, reproducing by seeds; stems 0.15 to 1 m high; introduced from Europe; in all provinces, in grainfields, cultivated fields, waste places, and along roadsides; this and some other smartweeds frequently have a dark blotch on the leaf surface; flowers first in June.

Green Smartweed *Polygonum scabrum* Moench

Annual, spreading by seeds; stems 0.35 to 1 m high; introduced from Europe; in all provinces, but more abundant in the Maritime Provinces and the Fraser Valley of British Columbia; in grainfields, cultivated fields, waste places, and along roadsides; flowers first appear in July.

Sheep Sorrel *Rumex acetosella* L.

Perennial, spreading by seeds and underground rootstocks; stems 15 to 30 cm high; introduced from Europe and Asia; occurs in every province, rare in the Prairie Provinces, and common in southern British Columbia and Eastern Canada; in meadows, pastures, and along roadsides; male and female flowers are on separate plants; flowers from June to October.

Curled Dock *Rumex crispus* L.

Perennial, spreading by seeds; stems up to 1 m high; introduced from Europe and Asia; in all provinces, but most abundant in the eastern provinces; occurs in meadows, pastures, waste places, and along roadsides; several other perennial docks resemble curled dock; flowers first in the middle of June.

Russian Pigweed *Axyris amaranthoides* L.

Annual, spreading by seeds; stems up to 1.35 m high; introduced from Asia; in every province except Newfoundland, but most common in the Prairie Provinces; in grainfields, gardens, farmyards, waste places, and along roadsides; flowers from June to August.

Lamb's-Quarters *Chenopodium album* L.

Annual, spreading by seeds; stems 0.35 to 2 m high; introduced from Europe; in all settled areas of Canada; in row crops, grainfields, gardens, and along roadsides; a number of closely related plants resembling lamb's-quarters are also widespread in Canada; flowers from June to September.

Kochia *Kochia scoparia* (L.) Schrader

Annual, spreading by seeds; stems 0.35 to 2 m high; introduced from Europe and Asia; in all provinces except Newfoundland, Prince Edward Island, and New Brunswick; most common in cultivated fields, waste places, and along roadsides in the Prairie Provinces and the interior of British Columbia; flowers from July to September.

Russian Thistle *Salsola pestifer* Nels.

Annual, spreading by seeds; stems a few centimetres to nearly 2 m high; introduced from Europe and Asia; abundant in the drier parts of the Prairie Provinces; after seeds mature, the nearly spherical plant breaks away near ground level, and the plant is rolled great distances by the wind; flowers first in early July.

Redroot Pigweed *Amaranthus retroflexus* L.

Annual, spreading by seeds; stems a few centimetres to 1.35 m high; native to North America; in all provinces except Newfoundland; grows in gardens, row crops, waste places, and along roadsides; plant has a characteristic red root, thus its common name; flowers from July to September.

Purslane *Portulaca oleracea* L.

Annual, reproducing by seeds; stem prostrate, stout, and fleshy; introduced from Europe and Asia; in every province except Newfoundland; most common in gardens but also in row crops, waste places, and along driveways; can produce seed without the flowers opening up; small yellow flowers appear first about the middle of July.

Mouse-Eared Chickweed *Cerastium vulgatum* L.

Perennial, forming patches; spreading both by seeds and by roots on prostrate stems; introduced from Europe; in every province, but most abundant on the Pacific coast and east of the Great Lakes; in lawns, gardens, pastures, cultivated fields, and along roadsides; flowers from early spring until late autumn.

White Cockle *Lychnis alba* Mill.

Biennial or short-lived perennial, spreading by seeds; stems up to 1.35 m high; native to Europe; in all provinces, but rare in the Prairie Provinces; in hayfields, grainfields, waste places, and along roadsides; male and female flowers are on separate plants; flowers from June to September.

Cow Cockle *Saponaria vaccaria* L.

Plate 23

Annual, spreading by seeds; stems 15 to 70 cm high; introduced from Europe and Asia; in every province except Prince Edward Island and Newfoundland; a serious weed only in grainfields of the Prairie Provinces, especially on fine textured soils; seeds are poisonous to animals; flowers from June to September.

Bladder Campion *Silene cucubalus* Wibel

Perennial, spreading mainly by seeds; forming clumps about 0.5 m high; introduced from Europe and Asia; in all provinces, but most common in Eastern Canada; in hayfields, cultivated fields, waste places, and along roadsides; flowers pinched at the open end can be popped against the palm; flowers from mid-June until September.

Night-Flowering Catchfly *Silene noctiflora* L.

Annual, spreading by seeds; stems usually up to 0.5 m high; introduced from Europe; in all provinces; sticky when squeezed between the fingers, whereas the superficially similar white cockle is not sticky; whitish flowers open only at night or on very dull days; flowers from June to September.

Corn Spurry *Spergula arvensis* L.

Annual, spreading by seeds; stems 15 to 45 cm high; introduced from Europe; in all provinces except Manitoba and Saskatchewan, but common only in southeastern Quebec, the Maritime Provinces, and southwestern British Columbia; in grainfields, row crops, gardens, and along roadsides; flowers from June to October.

Chickweed *Stellaria media* (L.) Vill.

Annual or winter annual, spreading by seeds; stems also root at their nodes; introduced from Europe; in all provinces, but most common in British Columbia and Eastern Canada; in grainfields, row crops, pastures, gardens, lawns, and waste places; flowers from early spring until late autumn.

Tall Buttercup *Ranunculus acris* L.

Perennial, reproducing by seeds; stems 0.5 to 1 m high; introduced from Europe; in all provinces, but most abundant in Eastern Canada; in pastures, hayfields, and along roadsides; usually not grazed by livestock, but if eaten, poisoning may result; flowers from late May until September.

Creeping Buttercup *Ranunculus repens* L.

Perennial with creeping shoots; frequently prostrate; introduced from Europe; in all provinces except Manitoba and Saskatchewan, but most common in the Maritime Provinces and coastal British Columbia; in pastures, lawns, and waste places; flowers from May to August.

Common Barberry *Berberis vulgaris* L.

Shrub, 1 to 3 m high, many stemmed from base; introduced from Europe; in all provinces, except Newfoundland, Saskatchewan, and Alberta; alternate host of stem rust of wheat, oats, barley, and some grasses; fruits bright red; flowers yellow, in loose drooping clusters; flowering late May to early June.

Yellow Rocket *Barbarea vulgaris* R. Br.

Biennial or perennial, spreading by seeds; stems 0.7 m high; introduced from Europe; in all provinces; particularly common in hayfields, pastures, and along roadsides in Eastern Canada; one of the first weeds to flower in the spring; large clusters of yellow flowers appear in May and June.

Bird Rape *Brassica campestris* L.

Annual or winter annual, spreading by seeds; stems 10 cm to 1 m high; introduced from Europe and Asia; in all provinces, but most common in the Maritime Provinces, adjacent Quebec, and coastal British Columbia; in cultivated fields, waste places, and along roadsides; forms fertile hybrids when crossed with the cultivated turnip and rape; flowers from April to June.

Shepherd's-Purse *Capsella bursa-pastoris* (L.) Medic.

Annual or winter annual, spreading by seeds; stems a few centimetres to 1 m high; introduced into North America from Europe before 1700 and now found in the settled areas of all provinces; in grainfields, row crops, gardens, and along roadsides; one of the most common weeds of Canada; flowers from early spring until late autumn.

Hare's-Ear Mustard *Conringia orientalis* (L.) Dumort.

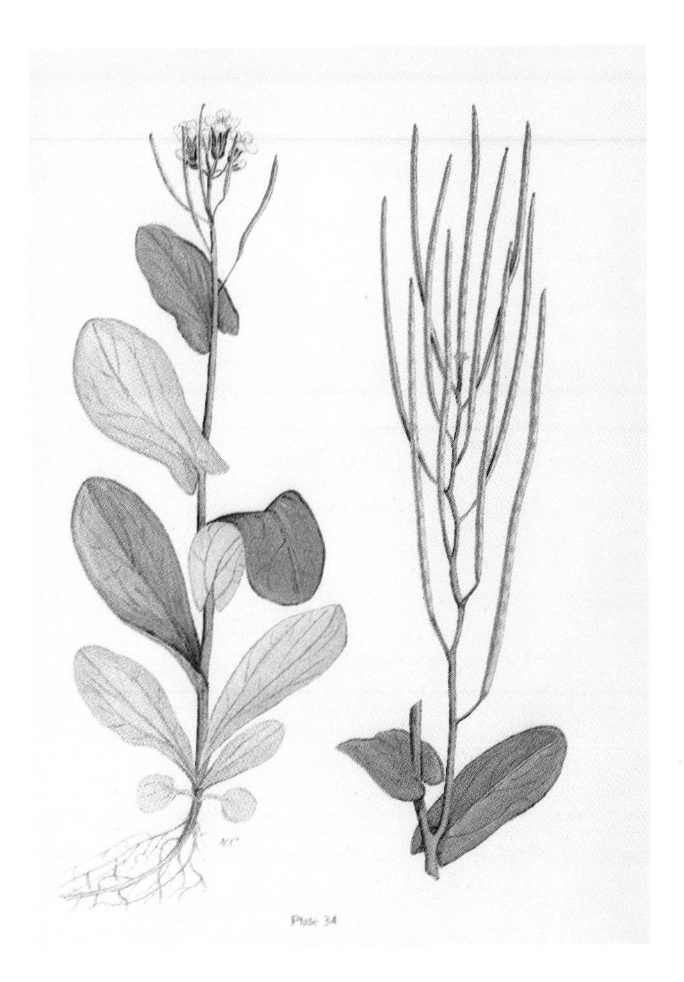

Annual or winter annual, spreading by seeds; stems 15 to 70 cm high; introduced from Europe; in all provinces, but most abundant in the Prairie Provinces; in fields, gardens, and along roadsides; seeds of this plant may cause poisoning when fed in grain; flowers from May until August.

Flixweed *Descurainia sophia* (L.) Webb

Annual or biennial, spreading by seeds; stems up to 1 m high; introduced from Europe; in all provinces; in grainfields, gardens, waste places, and along roadsides; one of the most abundant weeds of the Canadian prairies; flowers from May until July.

Wormseed Mustard *Erysimum cheiranthoides* L.

Annual or winter annual, spreading by seeds; stems a few centimetres to 1.5 m high; introduced from Europe; in all provinces, the District of Mackenzie, and the Yukon; in cultivated land, waste places, and also in native habitats; flowers from June until late autumn.

Tall Wormseed Mustard *Erysimum hieraciifolium* L.

Perennial, reproducing by seeds; stems 0.15 to 2 m high; introduced from Europe about 1950 and now known in Nova Scotia, Quebec, Ontario, and Saskatchewan; mainly along roadsides, in some areas forming dense stands; flowers first in mid-June.

Common Pepper-Grass *Lepidium densiflorum* Schrad.

Annual or winter annual, spreading by seeds; stems 10 to 70 cm high; native to North America; in all provinces; in cultivated fields, waste places, and along roadsides; flowers from June until August.

Ball Mustard *Neslia paniculata* (L.) Desv.

Annual, spreading by seeds; stems 0.35 to 1 m high; introduced from Europe as early as 1891 and now occurs in all provinces; particularly common in grainfields of the Prairie Provinces, along railway lines, and in waste places; flowers from June until September.

Wild Radish *Raphanus raphanistrum* L.

Annual or winter annual, spreading by seeds; stems 0.35 to 1 m high; introduced from Europe; most common in the Maritime Provinces, southern Vancouver Island, and the Fraser Valley of British Columbia; in grainfields, row crops, waste places, and along roadsides; flowers from May until July.

Wild Mustard *Sinapis arvensis* L.

Annual, reproducing by seeds; stems 0.35 to 1 m high; introduced from Eurasia; in every province; one of the most abundant weeds in grainfields of the Prairie Provinces; in grainfields, row crops, waste places, and along roadsides; distinguishable from other mustards by the presence of small downward-pointing hairs on the stem; flowers from June until autumn.

Tumble Mustard *Sisymbrium altissimum* L.

Annual or winter annual, spreading by seeds; stems up to 0.7 m high; introduced from Europe; in every province, but particularly abundant in the Prairie Provinces; in grainfields, grasslands, waste places, and along roadsides; at maturity, stems often break at the base, and the whole plant is blown by the wind; flowers from early spring until late summer.

Stinkweed *Thlaspi arvense* L.

Annual or winter annual, spreading by seeds; stems a few centimetres to 0.7 m high; introduced from Europe and Asia; in all provinces; in grainfields, hayfields, gardens, and waste places, and is most troublesome in the grain-growing areas of the Prairie Provinces; crushed leaves produce an unpleasant odor; white flowers present from early spring until late autumn.

Silvery Cinquefoil *Potentilla argentea* L.

Perennial, spreading by seeds; prostrate to semi-erect; introduced from Europe; in all provinces except Alberta, but most common in Ontario and western Quebec; in pastures, lawns, waste places, and along roadsides; flowers first appear in June.

Rough Cinquefoil *Potentilla norvegica* L.

Biennial, or short-lived perennial, spreading by seeds; stems 0.15 to 0.7 m high; native to North America; common in all provinces; in grainfields, hayfields, pastures, gardens, woods, and waste places; yellow flowers present mainly in June and July.

Sulfur Cinquefoil *Potentilla recta* L.

Perennial, spreading by seeds; stems 15 to 45 cm high; introduced from Europe; in all provinces, but most abundant in Ontario and western Quebec; in pastures, hayfields, waste places, and along roadsides; sulfur-yellow petals present mainly in July and August.

Choke Cherry *Prunus virginiana* L.

Large shrub or small tree up to 10 m high; native to North America; in all provinces and continental Northwest Territories; leaves and sometimes seeds of choke cherry and black cherry (*P. serotina* Ehrh.) can cause poisoning to livestock and humans; poisoning is by formation of hydrocyanic acid in stomachs; ripe fruits are not poisonous; fruits deep red to dark red-purple, rarely white or yellow; flowers from mid- to late May in south to mid-August in north.

Narrow-Leaved Meadowsweet *Spiraea alba* Du Roi

Perennial shrub; stems up to 1.35 m high; native to North America; occurs from southern Quebec to Alberta; in old fields, waste places, and along roadsides; clusters of white flowers are present from June until September.

Black Medick *Medicago lupulina* L.

Annual or winter annual, spreading by black seeds; mainly prostrate; introduced from Europe; in all provinces; in cultivated fields, pastures, waste places, and along roadsides; small clusters of yellow flowers are present from early spring until late autumn.

White Sweet-Clover *Melilotus alba* Desr.

Annual, spreading by seeds; stems up to 3 m high; introduced from Europe; in all provinces; mainly along roadsides and in waste places; sometimes sown for forage, cover crop, and green manuring; white flowers are present from May until October.

Yellow Sweet-Clover *Melilotus officinalis* (L.) Lam.

Annual, spreading by seeds; stems up to 2.5 m high; introduced from Europe; in all provinces; mainly along roadsides and in waste places; in many areas of Canada white and yellow sweet-clovers form solid stands along the edges of roadsides; yellow flowers are present from May until October.

Tufted Vetch *Vicia cracca* L.

Perennial, frequently twining about other plants; spreading by seeds and rootstocks; introduced from Europe; in all provinces, but most common in Eastern Canada; in meadows, pastures, gardens, waste places, grainfields, and row crops; flowers from early June until October.

Cypress Spurge *Euphorbia cyparissias* L.

Perennial, with underground rootstocks; stems usually about 0.35 m high; introduced from Europe; in all provinces except Alberta; milky juice may cause severe skin rashes in humans; plant is poisonous to most livestock; sometimes grown as an ornamental; flowers in June and July.

Leafy Spurge *Euphorbia esula* L.

Perennial, spreading by seeds and underground rootstocks; stems up to 1 m high; introduced from Europe; in all provinces except Newfoundland, but most common on prairie lands; milky juice may cause severe skin rashes in humans; plant is poisonous to most livestock; flowers in June and July.

Velvetleaf *Abutilon theophrasti* Medic.

Annual, spreading by seeds; stems up to 1.35 m high; introduced from India; in Prince Edward Island and from Quebec to Saskatchewan, but common only in southern Ontario; in cultivated fields, waste places, and along roadsides; flowers in August and September.

Common Mallow *Malva neglecta* Wallr.

Annual to short-lived perennial, spreading by seeds; stems prostrate to semi-erect; introduced from Europe; in all provinces except Saskatchewan, but most common in settled areas of Quebec, Ontario, and British Columbia; fruits form a disk about 1 cm in diameter; whitish to pale lilac petals present from May to October.

St. John's-Wort *Hypericum perforatum* L.

Perennial, spreading by seeds and by shoots from underground runners; stems 0.35 to 1 m high; introduced from Europe; in every province except the Prairie Provinces; in rangelands, pastures, meadows, waste places, and along roadsides; containing a toxic substance that affects white-haired animals when they are exposed to strong sunlight after eating the plant; flowers from June until September.

Poison-Ivy *Rhus radicans* L.

Woody perennial, spreading by seeds and sucker shoots; stem trailing along ground or climbing trees to 10 m high; native to North America; in all provinces except Newfoundland, but most common from Quebec City to the Great Lakes; juice from crushed portions of plant causes a characteristic skin blistering; fresh or dried juices carried on clothing, tools, dog fur, and such like, produce a rash when they contact the skin of sensitive individuals; small clusters of whitish green flowers are present from June to August; clusters of small, round, dull white berries first appear in mid-July. Another similar poisonous *Rhus* species, western poison-oak (*R. diversiloba* T. & G.), is confined in Canada to remote areas on the east coast of Vancouver Island and on some small adjacent islands.

Poison Sumac *Rhus vernix* L.

Shrub or small tree, 2 to 7 m high, spreading by whitish fruits; native to North America; wooded swamps in southern Ontario and southern Quebec; plant parts poisonous to the skin of most individuals; do not pick brightly colored autumn foliage; flowers from May to July.

Purple Loosestrife *Lythrum salicaria* L.

Perennial with underground rootstocks; stems up to 1 m high; a garden escape introduced from Europe; in all provinces except Saskatchewan, but most common in moist locations in Quebec, Ontario, and the Fraser Valley of British Columbia; flowers from June to September.

Yellow Evening-Primrose *Oenothera biennis* L.

Biennial, spreading by seeds; stems 0.7 to 2 m high; native to North America; in all provinces, but more common in the east than in the west; in pastures, waste places, and along roadsides; evening-primrose forms a flat rosette the first year; flowers from July to September.

Spotted Water-Hemlock *Cicuta maculata* L.

Perennial, 1 to 2 m high, spreading by seeds, globular rootstocks, and thickened storage roots; native to North America; in every province and north to the District of Mackenzie and the Yukon; in wet places; water-hemlocks are very poisonous to humans and livestock; white flowers are present from June to August. Two similar poisonous species are: western water-hemlock, *C. douglasii* (DC.) Coult. & Rose in British Columbia and northern water-hemlock, *C. virosa* L., in northern Canada.

Wild Carrot *Daucus carota* L.

Annual or biennial, spreading by seeds; stems up to 1 m high; introduced from Europe and Asia; in all provinces except Alberta, but most common in Quebec, Ontario, and the southern coast of British Columbia; forms fertile hybrids when crossed with the cultivated carrot; flowers first about the middle of July.

Wild Parsnip *Pastinaca sativa* L.

Biennial, spreading by seeds; stems up to 1 m high; introduced from Europe; in all provinces, but particularly common in eastern Ontario and Quebec; in pastures, hayfields, waste places, and along roadsides and riverbanks; contact with plant produces a skin irritation in some individuals; yellow flowers are present from July to October.

Common Milkweed *Asclepias syriaca* L.

Perennial, spreading by seeds and underground rootstocks; stems 0.70 to 1.35 m high; native to North America; in a wide range of habitats; occurs from Manitoba east to Prince Edward Island, but most common in southern Ontario and Quebec; plants contain a milky juice; flowers in June and July.

Field Bindweed *Convolvulus arvensis* L.

Perennial, spreading by seeds and underground roots; introduced from Europe; in every province except Newfoundland and Prince Edward Island, but most troublesome in the southern Prairie Provinces and long-settled areas of Ontario and Quebec; in cultivated land, grainfields, meadows, waste places, and along roadsides; stems often twine around other plants; flowers from June until September.

Dodders *Cuscuta* spp.

Annual parasitic flowering plants; spreading by seeds; introduced from Europe; some dodders occur in every province; dodders are parasitic on native, weedy, and cultivated plants that grow in a wide range of habitats; stems are orange or reddish and lack green chlorophyll; small white or cream-colored flowers are single or in clusters.

Blueweed *Echium vulgare* L.

Biennial, spreading by seeds; stems 0.35 to 1 m high; introduced from Europe; in all provinces, but rare and localized except in southern Ontario and adjacent Quebec; in permanent pastures, abandoned fields, meadows, waste places, and along roadsides; flowering from June to August.

Bluebur *Lappula echinata* Gilib.

Annual, or winter annual, spreading by seeds; stems up to 0.70 m high; introduced from Europe; in all provinces, but more abundant in the west than in the east; in grainfields, pastures, waste places, and along roadsides; small blue flowers are present in June and July.

American Dragonhead *Dracocephalum parviflorum* Nutt.

Annual or biennial, spreading by seeds; stems 0.35 to 1 m high; native to North America; mainly in the Prairie Provinces; in clover fields, grainfields, gardens, native grassland, and waste places; flowers from June to August.

Hemp-Nettle *Galeopsis tetrahit* L.

Annual, spreading by seeds; stems 0.15 to 1 m high; introduced from Europe and Asia; in all provinces; in grainfields, gardens, pastures, barnyards, waste places, and along roadsides; stems are covered with bristly hairs, which tend to penetrate the skin when the plant is handled; flowers from July to September.

Ground-Ivy *Glechoma hederacea* L.

Perennial, spreading mainly by creeping stems that root at nodes; introduced from Europe and Asia, probably as a ground-cover plant; in all provinces, but more common in Eastern Canada; flowers from late April to July.

Climbing Nightshade *Solanum dulcamara* L.

Woody climber; introduced from Europe; in all provinces except Saskatchewan and Alberta; in waste places and wood openings; fruit poisonous when eaten in quantity; purple flowers are present from May until September.

Yellow Toadflax *Linaria vulgaris* Mill.

Perennial, spreading by seeds and by underground roots; stems 15 to 67 cm high; introduced from Europe and Asia; in all provinces; grows in grasslands, cultivated fields, gardens, waste places, and along roadsides; flowers from June to October.

Common Mullein *Verbascum thapsus* L.

Biennial, spreading by seeds; stems 0.15 to 3.5 m high; introduced from Europe; in every province except Prince Edward Island, but rare in the Prairie Provinces; in pastures, waste places, and along roadsides; flowers from June to September.

Narrow-Leaved Plantain *Plantago lanceolata* L.

Perennial, spreading by seeds; flowering stems usually less than 35 cm high; introduced from Europe; in all provinces except the Prairie Provinces, but common in southwestern Ontario, locally common elsewhere, especially in the Maritime Provinces and southwestern British Columbia; pollen grains are a factor in early summer hay fever; flowers from June to October.

Broad-Leaved Plantain *Plantago major* L.

Perennial, spreading by seeds; flowering stems usually less than 20 cm high; introduced from Europe; abundant in the settled areas of all provinces; in lawns, pastures, meadows, cultivated land, waste places, and along roadsides; flowers from June to October.

Hoary Plantain *Plantago media* L.

Perennial, spreading by seeds; flowering stems usually less than 15 cm high; introduced from Europe; occurs from New Brunswick to Manitoba and in British Columbia, but common only locally; in lawns, waste places, and along roadsides; flowers from June to September.

Rugel's Plantain *Plantago rugelii* Decne.

Perennial, spreading by seeds; flowering stems usually less than 20 cm high; native to North America; in Nova Scotia, New Brunswick, and Ontario; grows along roadsides, shorelines, and in waste places; flowers from July to October.

Yarrow *Achillea millefolium* L.

Perennial, spreading by seeds and shallow roots; stems usually 35 to 70 cm high; native to North America; in all provinces; one of the most common weeds in Canada; in pastures, lawns, meadows, waste places, and along roadsides; flowers from June to August.

Common Ragweed *Ambrosia artemisiifolia* L.

Annual, spreading by seeds; stems 0.15 to 1 m high; native to North America; in all provinces, but commonest in southern Ontario and in southern Quebec as far east as Quebec City, and rare in the Maritime Provinces, the Prairie Provinces, and British Columbia; air-borne pollen grains of common ragweed are the most important cause of hay fever in eastern North America; it is estimated that there are 800,000 victims of ragweed pollen in Eastern Canada and 10 million in the United States; dairy products from cows that have grazed this plant have an objectionable odor and taste; flowers mainly in August and early September.

Giant Ragweed *Ambrosia trifida* L.

Annual, spreading by seeds; stems 0.15 to 3 m high; native to North America; in all provinces except Newfoundland, but particularly abundant in Manitoba, southern Ontario, and southwestern Quebec; pollen can cause hay fever; flowers from late July until early September.

Pearly Everlasting *Anaphalis margaritacea* (L.) Benth. & Hook.

Perennial, 10 to 90 cm high, spreading by seeds and underground rhizomes; native to North America; in all provinces, continental Northwest Territories, and the Yukon; common in disturbed habitats, particularly in the Maritime Provinces; easily dried for flower arrangements; flowering July to September.

Stinking Mayweed *Anthemis cotula* L.

Annual, spreading by seeds; stems 10 to 35 cm high; introduced from Europe; common in farmyards, waste places, and along roadsides in New Brunswick, Ontario, Quebec, and British Columbia, but rare elsewhere; flowers from June to October.

Common Burdock *Arctium minus* (Hill) Bernh.

Biennial, spreading by seeds; stems 0.70 to 2 m high; introduced from Europe; in all provinces, but most abundant in Eastern Canada; in farmyards, waste places, and along roadsides; mature flower heads form prickly burs that adhere to clothing and fur; rosette leaves resemble those of rhubarb; flowers from July to September.

Absinth *Artemisia absinthium* L.

Strongly aromatic perennial; stems up to 2 m high; introduced from Europe; in all provinces, but abundant only in the Prairie Provinces; in waste places, farmyards, pastures, cropland, and along roadsides; when eaten by cows, absinth causes a taint in dairy products; flowers from late July to September.

Plumeless Thistle *Carduus acanthoides* L.

Biennial, spreading by seeds; stems 0.20 to 2 m high; introduced from Europe, locally common in Quebec, Ontario, and British Columbia; in pastures, waste places, and along roadsides; flowers from June to September.

Nodding Thistle *Carduus nutans* L.

Biennial, spreading by seeds; stems 0.35 to 2 m high; introduced from Europe and Asia; in every province except Prince Edward Island and Alberta, but most common in pastures, rangeland, waste places, and along roadsides in Saskatchewan and Ontario; flowers from July to September.

Diffuse Knapweed *Centaurea diffusa* Lam.

Biennial to short-lived perennial; stems 0.70 to 1 m high; introduced from Europe and Asia; in southern Alberta and British Columbia, very common along roadsides and in dry rangelands of British Columbia; flowers from July to September.

Spotted Knapweed *Centaurea maculosa* Lam.

Biennial or short-lived perennial; stems 0.70 to 1 m high; introduced from Europe; in Nova Scotia, New Brunswick, Quebec, Ontario, and British Columbia; most abundant along roadsides and in dry rangeland of British Columbia and in pastures and along roadsides of Grey and Hastings counties of Ontario; flowers from June to October.

Black Knapweed *Centaurea nigra* L.

Perennial; stems 0.70 to 1.35 m high, coarsely branched with purple flowers and blackish or brownish involucre; introduced from Europe; most abundant in the Atlantic Provinces; occurs sporadically in Quebec, Ontario, and British Columbia; flowering July to September.

Ox-Eye Daisy *Chrysanthemum leucanthemum* L.

Perennial, spreading by seeds; stems 0.35 to 1 m high; introduced from Europe; in all provinces, but rare in Saskatchewan and Alberta; in meadows, pastures, waste places, hayfields, and along roadsides; when eaten by cattle this plant gives milk a disagreeable taste; flowers from June to August.

Chicory *Cichorium intybus* L.

Perennial, spreading by seeds; stems 0.35 to 2 m high; introduced from Europe; in all provinces, but abundant in Eastern Canada and southern British Columbia, and rare in the Prairie Provinces; in hayfields, waste places, and along roadsides; cows eating large quantities of this plant produce milk with a bitter flavor; flowers from July to September.

Canada Thistle *Cirsium arvense* (L.) Scop.

Perennial, spreading by seeds and underground rootstocks; stems 0.15 to 1.35 m high; introduced from Europe; in every province; in cultivated fields, meadows, pastures, waste places, and along roadsides; male and female flowers are on separate plants; flowers from June to October.

Bull Thistle *Cirsium vulgare* (Savi) Ten.

Biennial, spreading by seeds; stems 0.35 to 2 m high; introduced from Europe and Asia; in all provinces, but most common in Eastern Canada and southern British Columbia; in pastures, waste places, and along roadsides; flowers from June until September.

Smooth Hawk's-Beard *Crepis capillaris* (L.) Wallr.

Perennial, spreading by seeds; 0.15 to 1 m high; introduced from Europe; common on Vancouver Island and the adjacent mainland; in meadows, pastures, waste places, and along roadsides; flowers from May to November.

Canada Fleabane *Erigeron canadensis* L.

Annual or winter annual, spreading by seeds; stems a few centimetres to 2 m high; native to North America; in all provinces, but less common in the Maritime Provinces; in cultivated fields, pastures, meadows, waste places, and along roadsides; flowers from July to October.

Philadelphia Fleabane *Erigeron philadelphicus* L.

Perennial, reproducing by seeds, stolons, and offsets; stems 0.35 to 1 m high; native to North America; in all provinces, but less common in the Prairie Provinces and the Maritime Provinces; in hayfields, pastures, waste places, and along roadsides and riverbanks; flowers from May to September.

Rough Fleabane *Erigeron strigosus* Muhl.

Annual or biennial; stems 0.70 to 1.35 m high; native to North America; in all provinces, but less common in the Prairie Provinces; in pastures, hayfields, waste places, and along roadsides; flowers from June to October.

Hairy Galinsoga *Galinsoga ciliata* (Raf.) Blake

Annual, spreading by seeds; stems 15 to 70 cm high; introduced from South America; in all provinces except Newfoundland, but most abundant in Quebec, Ontario, and British Columbia; often grows close to buildings; flowers from June to October.

Gumweed *Grindelia squarrosa* (Pursh) Dunal

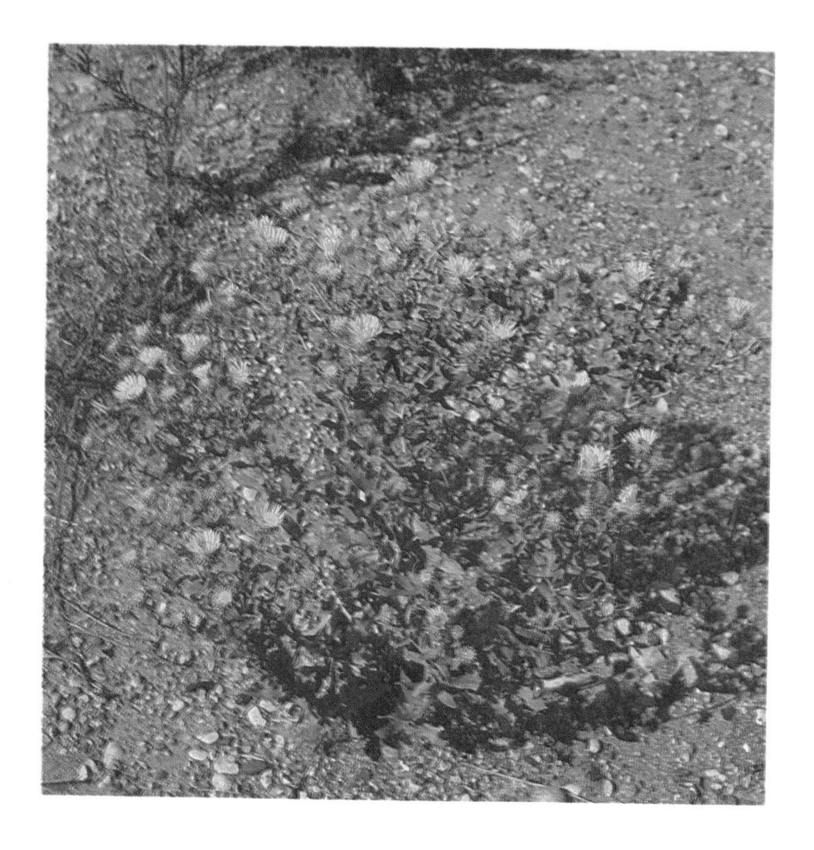

Perennial, reproducing by seeds; stems 0.10 to 1 m high; native to North America; most common in prairie areas from western Ontario to British Columbia and sporadic in the rest of Ontario, in Quebec, and in continental Northwest Territories; flowering involucre sticky; flowering from July to September.

Orange Hawkweed *Hieracium aurantiacum* L.

Perennial, stems 15 to 70 cm high; introduced from Europe; very abundant in Ontario, Quebec, and parts of the Maritime Provinces; in old fields, pastures, waste places, and along roadsides; flowers from June to October.

King Devil Hawkweed *Hieracium florentinum* All.

Perennial with short rootstocks; stems 15 to 50 cm high; introduced from Europe; abundant in Ontario and western Quebec; in pastures, lawns, waste places, and along roadsides; flowers from May to August.

Spotted Cat's-Ear *Hypochoeris radicata* L.

Perennial; stems 15 to 70 cm high; introduced from Europe; common only on Vancouver Island and the Queen Charlotte Islands of British Columbia and on the adjacent mainland; in pastures, waste places, and along roadsides; flowers from May to October.

Povertyweed *Iva axillaris* Pursh

Persistent perennial spreading by seeds and underground rootstocks; stems 15 to 20 cm high; native to the western prairies; common in the Prairie Provinces and less common in the interior of British Columbia; pollen can cause hay fever when plant is abundant; flowers from June to August.

False Ragweed *Iva xanthifolia* Nutt.

Annual; stems 1 to 2.5 m high; native to the western prairies; common in the Prairie Provinces and rare in British Columbia, Ontario, and Quebec; in cultivated land, waste land, and gardens; contact with leaves causes a rash in some people, and the pollen is an important cause of hay fever; flowers mainly in late August and early September.

Blue Lettuce *Lactuca pulchella* (Pursh) DC.

Perennial with deep rootstocks; stems up to 1 m high; native to western North America; mostly in open prairie, waste places, along roadsides, and in irrigated fields of the Prairie Provinces and British Columbia; flowers from June to August.

Prickly Lettuce *Lactuca scariola* L.

Annual or winter annual, spreading by seeds; stems 0.35 to 2 m high; introduced from Europe; in every province except Newfoundland, but most abundant in southern Ontario and the Prairie Provinces; in cultivated land, waste places, and along roadsides; flowers from mid-July to mid-September.

Fall Hawkbit *Leontodon autumnalis* L.

Perennial, spreading by seeds; stems 10 to 70 cm high; introduced from Europe and Asia; most common in the Maritime Provinces and southeastern Quebec; the dandelion, a close relative, has a hollow leafless stem bearing a single flower; flowering from early July to late September.

Scentless Chamomile *Matricaria maritima* L.

Annual to short-lived perennial reproducing by seed; 1.5 cm to 1 m high; introduced from Europe; in all provinces; most common in the Maritime Provinces and the adjacent part of Quebec and in the Prairie Provinces; crushed leaves of stinking mayweed have a strong odor, whereas those of scentless chamomile are practically odorless; flowers from late June to September.

Pineappleweed *Matricaria matricarioides* (Less.) Porter

Annual, spreading by seeds; stems 2 to 20 cm high; native to western North America; in all provinces and around settlements in the Yukon Territory and the Northwest Territories; in gardens, waste places, along roadsides, and on trampled ground; crushed leaves produce a pineapple odor; flowers from June to October.

Black-Eyed Susan *Rudbeckia hirta* L.

Perennial, spreading by seeds; stems 0.35 to 1 m high; native to North America; in all provinces; locally common; in hayfields, pastures, rangeland, waste places, and along roadsides; flowers from June to October.

Tansy Ragwort *Senecio jacobaea* L.

Biennial, or short-lived perennial; stems 0.35 to 1 m high; introduced from Europe; in all provinces on the Atlantic seaboard, and on Vancouver Island and the adjacent mainland in British Columbia; in pastures, hayfields, waste places, and along roadsides; can cause poisoning of cattle and horses; flowers in late July and August.

Canada Goldenrod *Solidago canadensis* L.

Perennial; stems 0.35 to 1.5 m high; native to North America; in all provinces; in waste places, unused farmland, and along roadsides; goldenrods as a group are common throughout Canada; flowering from late July to October.

Perennial Sow-Thistle *Sonchus arvensis* L.

Perennial, spreading by underground rootstocks; stems 0.35 to 1.5 m high; introduced from Europe and Asia; in all provinces; grows in grainfields, row crops, waste ground, and along roadsides; flowers from June to September.

Spiny Annual Sow-Thistle *Sonchus asper* (L.) Hill

Annual, spreading by seeds; stems 0.35 to 1.35 m high; introduced from Europe; in all provinces, but most abundant in Ontario, Quebec, and British Columbia; in cultivated fields, gardens, waste places, and along roadsides; flowers from June to September.

Annual Sow-Thistle *Sonchus oleraceus* L.

Annual, spreading by seeds; stems 0.35 to 1.5 m high; introduced from Europe; in all provinces; grows in gardens, row crops, waste places, and along roadsides; flowers from June until October.

Dandelion *Taraxacum officinale* Weber

Perennial, spreading by seeds; stems a few centimetres to 35 cm high; introduced from Europe; one of the most common weeds in all settled areas of Canada; in pastures, hayfields, cultivated land, lawns, waste places, and along roadsides; flowers from April to August.

Goat's-Beard *Tragopogon dubius* Scop.

Biennial to perennial, spreading by seeds; stems 15 to 70 cm high; introduced from Europe; common in the Prairie Provinces and fairly abundant in British Columbia, Ontario, and Quebec; in pastures, prairie, hayfields, waste places, and along roadsides; flowers from late May until July.

Meadow Goat's-Beard *Tragopogon pratensis* L.

Biennial to perennial, spreading by seeds; stems 0.70 to 1.35 m high; introduced from Europe; in all provinces except Newfoundland, but most common in Eastern Canada; in pastures, hayfields, waste places, and along roadsides; flowers in June and July.